SCIENCE IN AND AROUND YOU

PAMPHLET 1: AN INTRODUCTION TO SCIENCE

Copywrite © ALAN JOSEPH BAUER, Ph.D.

1

Table of Contents

Dedication

This work is dedicated to the memory of Rabbi Shmuel David Marzel, of blessed memory. He always saw in others their potential and treated them as if they had already reached it.

Background

The purpose of this pamphlet is to introduce science to audiences that generally are not exposed to it, whether they be non-science professionals or students just starting their studies. The book offers introductory discussions in the fields of biology, chemistry and physics. There is no attempt to produce a textbook—there are plenty of very good ones in the world. Rather, the goal is to show that truly amazing "science" is in you and all around you. The pamphlet begins with biological phenomena occurring right now inside your body and continues with the chemicals that are the basis for all of the things we have in our world. The brief journey ends with a quick look at physics and the laws that govern the behavior of phenomena in our world.

This book is meant for

*Professionals who wish to strengthen their knowledge and background in the sciences;

*Students who wishes to have a good reference for scientific information;

Non-scientists who wish to have a better idea of how the world around them works; and,

Anyone not mentioned above!

There is no attempt to be complete here. There are massive textbooks on each and every area of science. The goal here is to make an introduction. Did you know that you have tens of trillions of cells in your body? Did you know that the SR-71 needed titanium because if it had been made out of a standard aluminum alloy, the plane would have melted at Mach 3? Do you know that much of the electricity we use is made by converting one form of energy—such as water falling in a dam—into another form, namely electricity which we use for all of our devices and appliances (and today, many cars)? I hope that this book encourages more interest and delving into the wonders of the scientific world around us.

This volume is a general introduction to the sciences. Following pamphlets will deal with specific subjects, the first to be fission and the release and use of nuclear energy.

I hope that you enjoy this pamphlet as much as I have truly enjoyed writing it.

By way of background: I was born in Chicago in 1965 and grew up in the suburb of Wilmette, Illinois. My parents had fled Nazi Germany as children prior to the war, one via New York and the other by way of Sydney, Australia, and they met in Chicago. After studying biochemistry at Harvard and University of Wisconsin-Madison, I came to Israel on a U.S. Fulbright Postdoctoral Fellowship in 1992. My wife and I were married in 1994 and we have 4 wonderful boys. I am the inventor or co-inventor on 10 US patents.

When I was an undergraduate, many students who had come to study science soon found themselves studying other fields. The major reason my friends left science was that they found it boring and somewhat heavy. The world around us—including ourselves—is amazing. I hope that the excitement and beauty that I will try to present herewith comes through.

Jerusalem

30 April 2020

Cover art, Eagle Lake, California 96150 and photograph of Bay Bridge on page 44 generously provided by Mr. Binyamin B. Bauer

All images are licensed from shutterstock.com

I wish to thank my wonderful family for their love, support and constant curiosity.

I would like thank Mr. Bob Hendricks for fruitful discussions and encouragement.

Chapter 1: Biology

Biology is the studying of living things, from single-cell bacteria to humans, who have tens of trillion cells per person. Bacteria are living; viruses are not (1). Why are viruses not considered "alive"? They are not considered living because they reproduce only through the cells of a host—they cannot reproduce on their own but rather they highjack the machinery of the cells they infect in order to make more viral copies before exploding their host cell and spreading new viral particles to other cells. Below we'll talk about mechanisms the body has to fight viruses and other disease agents.

Before we get into a more detailed discussion, let's just put to paper some amazing statistics related to human beings. Get to know yourself:

*Estimated average number of cells in the human body: 37.2 trillion (2)

*Length of DNA in a person. Not all cells have DNA, for example red blood cells do not so as to carry as much oxygen as possible. Let's assume 10 trillion cells have DNA, and the accepted length of DNA is 2 meters per cell (3). That would come to well over 20,000 round-trips to the moon and back for the DNA in your body (3)

*Total length of blood vessels (veins and arteries) in your body: 60,000 miles (4)

*Alveoli, the 700 million little air sacs in the lungs that allow for oxygen to get into the blood, if spread out would cover a surface area of 750 square feet (5)

*Number of genes in the DNA: 20,000 spread over 46 chromosomes (6)

*Fastest enzyme (an enzyme is a protein that can perform one or more chemical reactions): carbonic anhydrase at 1 million reactions per second (7)

*Estimated number of enzymatic reactions per second in the body: 10^{27}—more than the number of stars in the universe (8)

*Number of heart beats from birth to 80 years old: over 3 billion (9)

The reason I start with these statistics in this section is that we often don't know—or appreciate—how amazing our bodies and all of their contents truly are. I always remain in awe when I think of enzymes performing hundreds or thousands of reactions per second or the amazing processes that occur in our various organs, without our being aware. Our bodies are incredible; it's worth knowing a little more about them.

Let's talk a little about the start of a human life. Eggs and sperm have half quantities of DNA (10), so that when one of each joins the opposite, the resulting cell has a full set of genes—half from the sperm and the other half from the egg. When a sperm enters an egg and successfully fertilizes the latter, a series of chemical reactions occur that prevents any further sperm penetration (11). If more sperm were to enter, an incorrect amount of DNA could cause improper growth of the dividing cells. The initial cell doubles several times, each cell identical to all of the others. After about 4 days of cell multiplication, differentiation occurs in which the different cells have different genes turned on or off (12). The result is that some cells go off and form bones while others form skin, while still others differentiate into the various internal organs. The control and communication between the cells are not fully understood, though mechanisms and chemicals involved in the process have been identified. Clearly there is exquisite coordination so as to have one left arm and one right arm and have the heart placed in the right location, etc. As the cells are doubling, not all are needed, and some cells are programmed to die in a process known as apoptosis (13).

Let's now look a little more into the details of a few of our senses.

Vision

We have all experienced being in a dark place and then suddenly having a bright light turned on. What happens? We cover our eyes or squint and after a few seconds everything is okay. We have also all had the experience of suddenly having the lights go out and for a few seconds we see nothing, and then later we can identify the objects in black and white around us and navigate them safely. These two phenomena are related to the two types of photoreceptor cells we have in our eyes: rods and cones (14). Rods are very sensitive to low light and as such provide us with our nighttime vision. Cones are optimized for high light conditions. When we go from dark to light or vice versa, we are essentially changing which sets of cells the eyes use in order to record data to be sent to the brain for processing. The retina of our eye contains about 120 million rod cells and 6 million cone cells (14). These cells are attached to nerves that allow transfer of images to processing centers in the brain. And yes, the images are inverted in the eye and then made right in the brain. I had a high school biology teacher who was involved in medical experiments in college in order to earn some money (15). He was given glasses that turned everything upside down. He could not take the glasses off. He said that the first few days were awful—everything was upside down. He could not move around much and he had headaches. After a few days, everything returned to normal. But what? When he would take off the glasses, everything was upside down!

Below is a schematic view of the "electromagnetic spectrum". Where you see the rainbow in the middle is the sliver of this spectrum that our eyes can see. As the same teacher taught us in biology: ROY G. BIV: Red-Orange-Yellow-Green-Blue-Indigo-Violet. As you can see from the chart, light in general and visible light specifically are part of a much larger group of phenomena that include x-rays, microwaves, radio and gamma waves. The only difference between the light that we can see and x-rays used at the hospital is the wavelength of the waves involved and their associated energies. The shorter the wavelength, the more energy the wave has associated with it. We cannot see x-rays but they are essentially identical in their properties to white light, which itself can be broken down by wavelength in a prism to the visible spectrum, each color having a different specific wavelength and thus moving differently through a prism—and thus separating into a rainbow. The human eye is capable of distinguishing between ten million unique colors (16). Notice that as the wavelength gets shorter, the energy gets stronger.

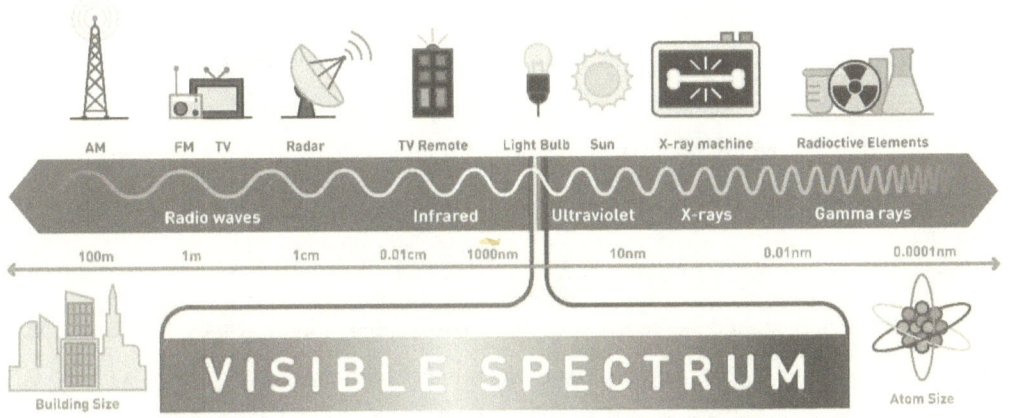

Many of us have need for either corrective lenses (glasses or contact lenses) or corrective surgery. Our eyes focus light received onto a certain point in the back of the eye. When the eyes function perfectly, we see the images around us as they appear. When there is a change in the shape of the lens in the eye, the focused light entering the eye does not reach this point

(17). Glasses change the approaching light in order to fix this problem and turn blurry images into sharper ones. Below shows the change in lens focus ("accommodation") of an object as a function of its distance from the eye. Note that the image is inverted as it is focused onto the retina for conversion to neural signals to be sent to the brain.

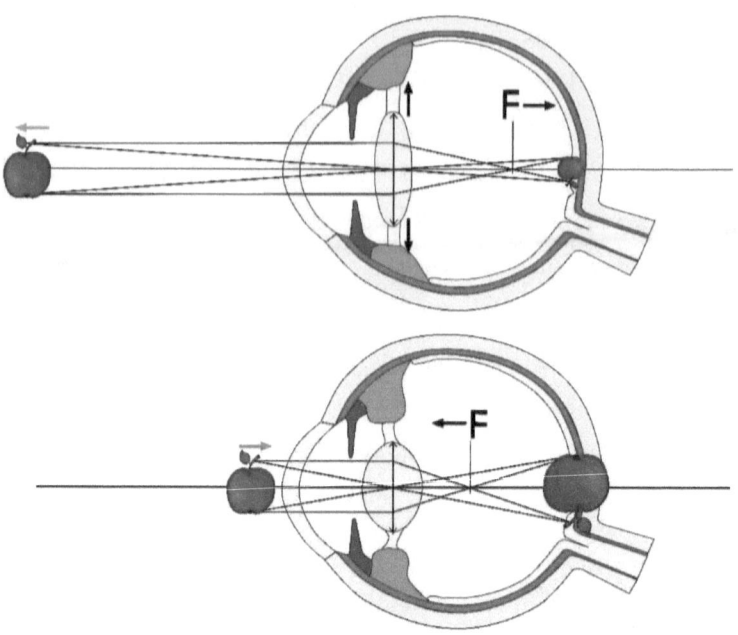

The teacher with the glasses above had us do an experiment. He put up a slide of a red square and told us to look at it for a minute. He then removed the image and everyone saw a green square where it had been though in reality there was nothing on the screen. The nerves telling the brain for a long time, "Red, red, red, red..." eventually wore out the response and when the square was removed, the brain had a back-reaction thinking that it was seeing a green square where the red square had been (18).

Hearing

Our ears are also designed to receive waves, these being sound waves. Our ears have tremendous dynamic range of hearing. Audible sound is generally between 20 and 20,000 Hertz in frequency and the strength of sound in is between 0 and 140 decibels that we can hear (19). Ultrasound, as it sounds, refers to sound waves of frequencies higher than 20,000 Hz, and yes, there are infrasounds of less than 20 Hz. Very loud music for extended periods of time can damage hearing, and one should be careful not to expose the ears to excessive levels of noise or music.

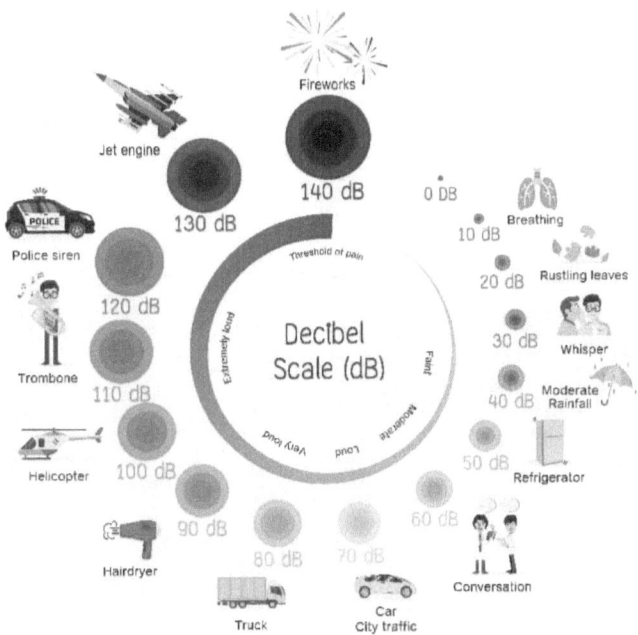

We can hear from the quiet whisper of a friend to the massive roar of a jet engine. The two ears, like our two eyes, provide stereo inputs which the brain averages in order to come up with a sound and determine its location. The sound waves enter the ear and hit the ear drum which vibrates. Those vibrations are passed into the inner ear where the vibrations cause motion of thousands of tiny hair cells. These motions are converted into an electrical signal that is sent to the brain via the hearing nerve. And in this way the brain figures out what was heard. As we get older, we may lose part of our range of sounds that we can hear. Thus, one of our boys has great glee to play sounds that he can hear but I cannot.

Taste and Smell

Whereas the eyes and ears receive waves and translate them into sight and sound, taste and smell are based on actual chemicals and their interaction with protein receptors designed to bind them and report back to the brain what is present ([20]). People know of "taste buds" that sit on the tongue. We have several thousand taste bud cells. It was originally thought that the different tastes were located on different parts of the tongue. That theory was discarded and it is accepted today that all tastes can be identified throughout the tongue. Below is a view of the cells involved in taste and the neurons on the other end that help transfer the information gleaned by the interaction of the cells with food to the brain. We have on our tongue cells with protein receptors, each cell type adapted to detect one type of taste: salty, sweet, sour, bitter, and umami.

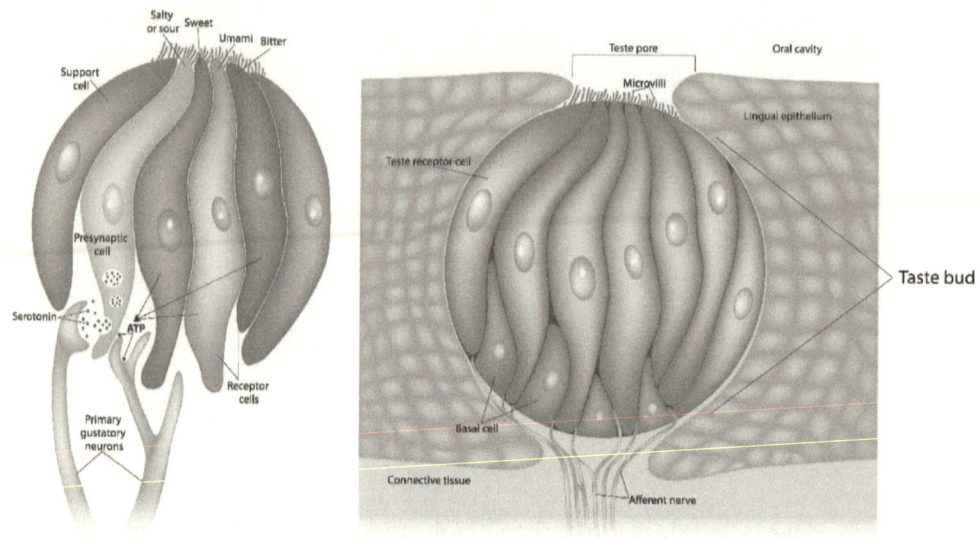

The different tastes are divided on the tongue as shown in the figure above. It is a combination of taste, smell, and texture that gives us a feeling for the foods we eat. The same biology teacher who did the experiment above with the upside-down glasses also participated in an experiment in which they put a divider between his mouth and nose. He would eat one type of food (without seeing it), say a steak, but to the nose they would give a different smell, say peanut butter. The brain had to figure out what it was eating when the smell did not match up with the taste and texture in the mouth.

The taste bud receptors are proteins (21), and as one knows that boiling water kills bacteria, high temperatures can denature (cause loss of structure to) proteins (22). So when we eat hot food, the change in protein structure of a thermoreceptor sends a warning to the brain that the food is too hot (23). We either remove the food or add cold water to reduce the temperature. As is well-known, hot peppers give us the feeling of "hot" though they are at room temperature. Capsaicin, the active ingredient in hot peppers, binds to the same proteins and causes the change in structure the proteins undergo when they get overheated—the result is that the brain gets a HOT!!! warning though there is nothing actually hot in the mouth (24). Because this chemical binds to the proteins, it takes a while to remove it and get rid of that hot feeling. Mint, which gives a cool feeling in the mouth, also has a chemical, menthol, that binds to these proteins and turns them off, thus causing the proteins to the brain that it's cool in the mouth (25). Again, the mint is at room or mouth temperature, but its interaction with the protein receptors in the mouth give the brain a signal that is taken to mean cool.

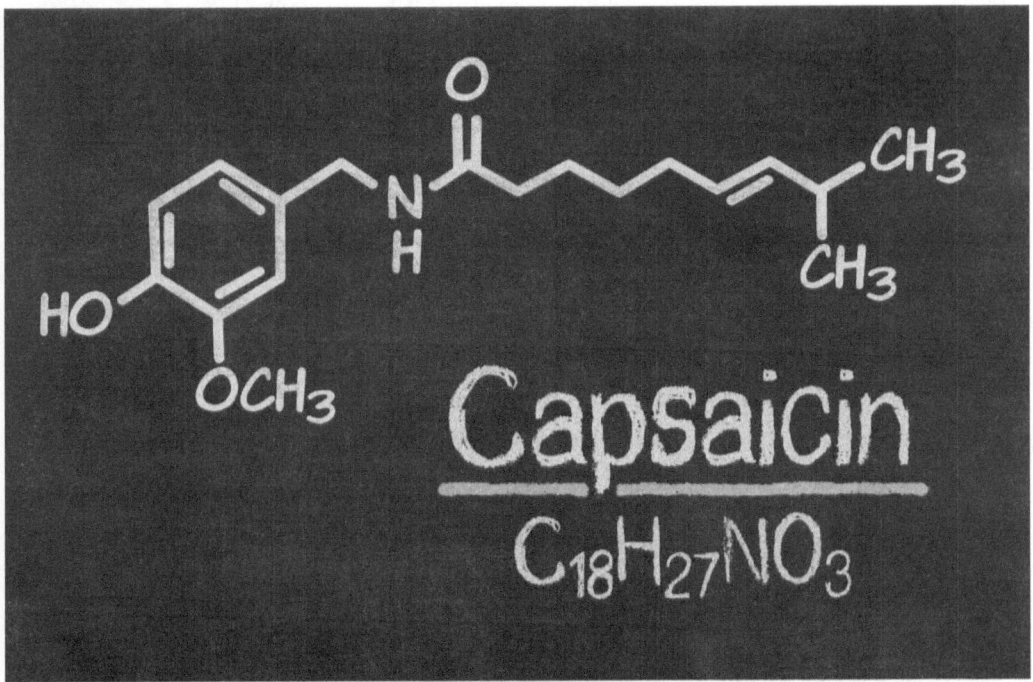

Chemicals binding to proteins or other biological entities is the basis of the drugs used to treat illness. Capsaicin is an example of a molecule binding to something in the body and causing a reaction, in this case one that can be either pleasant or horribly painful.

Like taste, smell is based on chemicals entering the nose and interacting with proteins that send along to the brain information as to what was smelled. The nose has approximately 400 different types of scent receptors and can distinguish through combinations of different odors 1 trillion unique smells (26). As an example, chocolate is comprised of nearly 600 chemicals including several dozen "aroma active" molecules that when the different receptors in the nose bind to them, the information is given to the brain which in turn concludes that it is smelling dark chocolate (27). Initially, it was thought that the nose could detect 10,000 unique scents but later research raised the number to a trillion.

A different system that works internally but is still subject to external stimuli is the immune response. Our body is able to produce in excess of ten *billion* different antibodies with only a handful of genes that can be changed and combined to produce different antibodies, a subset of which can bind to different potential targets like bacteria and viruses (28). The B-cells in the blood produce antibodies that are either released into the plasma to bind pathogens or have the antibodies tethered on the outside of the B-cells (29). When an antibody binds to a foreign agent, it can either interfere with the agent's ability to act against us and/or alert other parts of the immune system that the bound target is foreign and must be destroyed. The cells that produce the antibodies used against a given pathogen remain extant so that the antibodies will be available if they should again face the foreign antigen to which they successfully bind (29). Antibodies tend to be very specific in their binding of the epitope (the region on the antigen where the antibody binds) of an antigen, and these interactions generally can only be disrupted by extremes of salt, detergents, pH or the like (30).

So much has recently been written about Covid-19 and antibodies that I will not add anything here as there is a debate as to whether the antibodies for this virus can provide protection from future infections (31). I was once at a chemistry conference in Haifa. A very smart fellow from England presented his data on how the immune system identifies cells infected with viruses (32). Viruses enter cells, where they commandeer the host's protein production equipment in order to produce their own proteins provided by the nucleic acids they harbor. Those proteins are different than the normal proteins that would be found in say a human cell. A normal part of a protein's lifecycle is to be broken down into smaller parts (known as peptides) and finally back into amino acids for re-use in new proteins, either of the same or different variety. What this fellow pointed out was that these small peptides can be exported from the cell and placed on the outer surface of the cell. Think of the cell as if it is hanging its laundry outside the window to dry. Certain white blood T-cells contact the outer surface and due to the unnatural peptides present—those from the virus—these cells know that something is wrong and signal the target cell for death. These differences in correct and incorrect structure presented on the outside of the cell are slight, yet those small differences are enough to allow the immune system to identify a compromised cell and destroy it before it can release more virus particles to further infect the host.

The idea of self and non-self is extremely relevant in organ transplants. As is well-known, the body often rejects transplants because the immune system a) recognizes that the new thing present is not from the body and b) it is trained to identify and destroy foreign bodies (33).

People who undergo organ transplants often spend the rest of their lives on medication, including cyclosporin, an eleven-amino acid cyclic peptide produced by a fungus (34). A Harvard professor of biochemistry, after a fifty-year career at the university, said that he would in his retirement pursue one project only. He said that the body should reject a fetus, but it does not. To quote Professor Jack Strominger himself: "Why isn't the fetus rejected as if it's a foreign graft?" he said. "The solution to that problem is profound." (35)

In my first biochemistry course at Harvard, we learned about something called "the lactose operon" (36). Lactose, commonly known as milk sugar (37), is comprised of glucose and galactose bonded to one another. Cells have as their common energy currency glucose: enzymes work through glycolysis and the Krebs cycle (38) to break down glucose and realize the energy in its bonds in order to produce energy-rich molecules for the cell's normal metabolism. Thus, lactose "as is" is somewhat useless to a cell—like having euros where they only accept dollars. So, the cell must break the bond connecting glucose and galactose and later convert galactose to glucose in order to realize the maximum energy from this sugar. One of the enzymes performing that latter conversion was the basis of my Ph.D. studies in Madison. If the cell is not presented with lactose, there is no reason to waste cellular energy and resources to make the enzymes involved in its metabolism, so their normal cellular concentrations are low. On the other hand, if lactose is present, it would be a waste to lose it because the key enzymes needed to break down this sugar were not around in sufficient numbers. So, in the bacterium *E. coli* the genes for the relevant galactose-associated enzymes are grouped together and under control of a system that allows expression of those genes in response to lactose being present in the cell. Lactose entering the cell binds to a protein controller that keeps the genes related to lactose breakdown turned off. When lactose is bound, the protein changes structure and allows for the expression of the genes coding for the enzymes needed to metabolize lactose into glucose for energy realization. As lactose concentration drops due to its breakdown, less of the gene control protein is bound up with lactose. The protein again shuts down the genes in parallel to the completion of lactose processing. The cell cannot afford to waste energy: the lactose operon is only turned on when the relevant genes in the DNA are needed to realize energy from this particular sugar. In the figure below, lactose is called the "inducer" as it induces the system to go from off to on to allow the enzyme RNA polymerase to copy the DNA of the genes involved in lactose metabolism for assembly of the three requisite enzymes as labelled at the base of the figure.

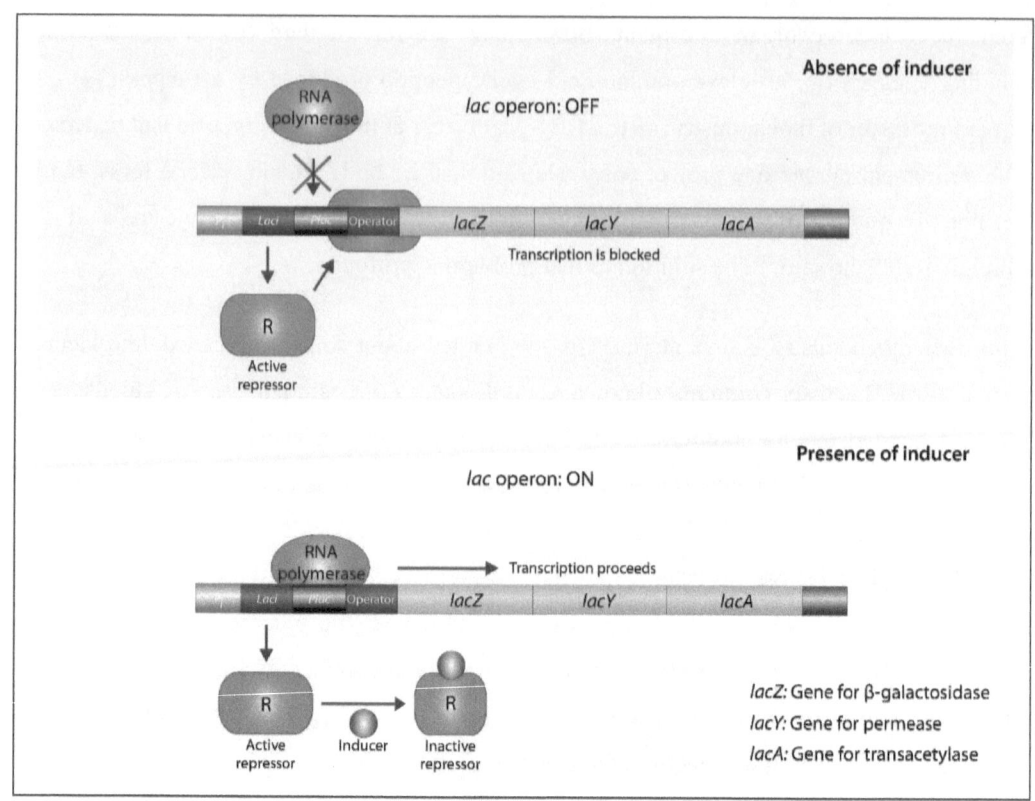

Absence of inducer

lac operon: OFF

RNA polymerase

Loci · *Plac* · Operator · lacZ · lacY · lacA

Transcription is blocked

R
Active repressor

Presence of inducer

lac operon: ON

RNA polymerase → Transcription proceeds

Loci · *Plac* · Operator · lacZ · lacY · lacA

R
Active repressor → Inducer → R Inactive repressor

lacZ: Gene for β-galactosidase
lacY: Gene for permease
lacA: Gene for transacetylase

There is a phenomenon we all know referred to as "fight or flight response" (39). Picture putting your foot into the street and then looking to your left. A huge bus in barreling down on you and you very quickly lift your foot and get back on the sidewalk just as the bus passes you by. Your heart is racing, you might even be breathing hard and feel your mouth is dry. What just happened? What you experienced was a very complex set of biological events that involved a great deal of your internal systems to save you from danger.

Inputs from the eyes and ears reach the brain that instantly recognizes danger. The brain causes the pituitary gland to release the hormone ATCH which in turn causes the release of epinephrine which we commonly call adrenaline. This molecule can bind to the outside of heart cells (it would take too long to get it into the cells themselves) and activate increased sugar metabolism and thus a higher heart rate. The overall effect is to provide the body with an instant energy boost to allow for a very fast response to a perceived danger. Other systems, as shown in the figure below, are also brought into the response. Just as quickly as it is turned on so it is turned off and generally a few minutes after a fright, our pulse and other indicators are back to normal. It takes a lot longer to explain what happens in the body when we experience a fight or flight response than it takes the body to go through all of the complex biochemical steps to get us to respond to a perceived threat.

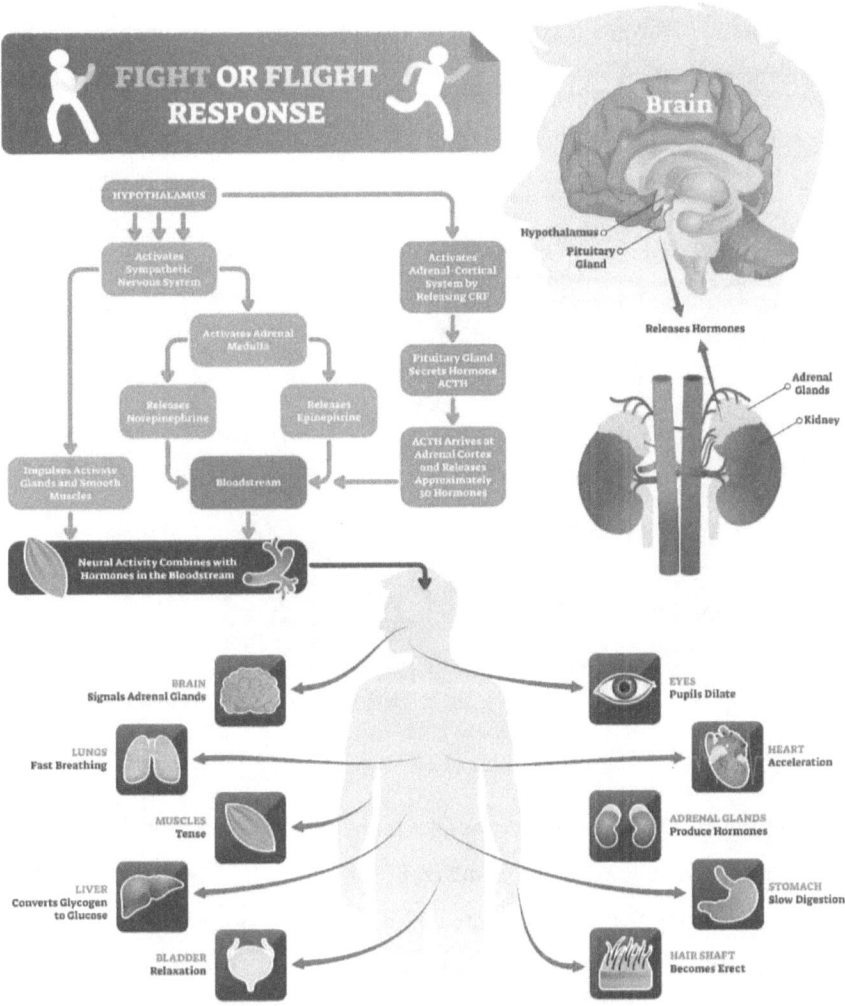

Plants

We have spoken about the body and a little about bacteria. What about plants? Have you ever noticed flowers opening and closing according to the time of day? Have you seen a plant turn towards the sun? What about seeing a plant growing upside down right itself to grow upright. Tropisms are processes whereby plants can respond to various environmental stimuli—light, water, gravity, chemicals, and more (40). A plant can selectively supply different levels of hormones—auxins—to different sides so as to allow one to grow more than the other and thus let the plant tilt towards the sun or a water source or the like (41). Anyone who has had a pipe ruptured by roots knows that roots know how to find water sources via "moisture gradients" (42). Apparently, one way they identify potential water is via the vibrations of water in a pipe (42). In the schematic below, the plant is shown to apply growth hormones unequally so as to make the plant grow straight as defined by the gravitational field being experienced by the plant.

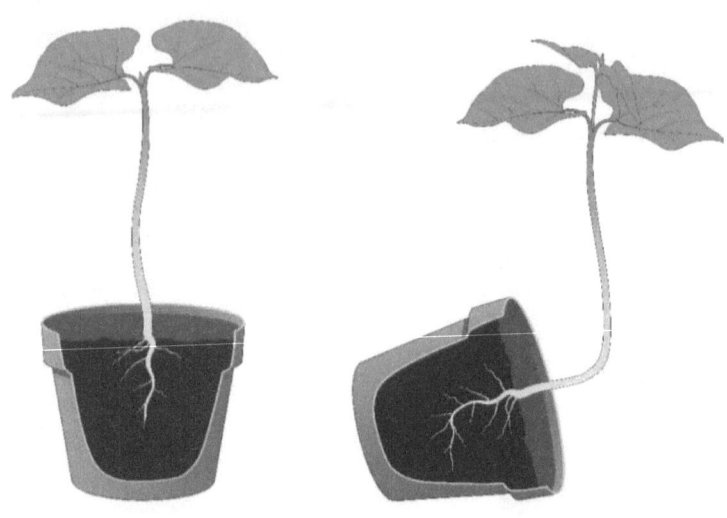

As mentioned in the introduction, this is not a textbook, and no effort has been made to give a complete view of the exceptionally broad and complex biological world. The goal has been to show some of the amazing features associated with biological systems, including those in our own bodies.

<u>Illness</u>

Above I alluded to the behavior of viruses in host cells. As is well-known, AIDS is related to the Human Immunodeficiency Viruses (HIV, 43). HIV, unlike Covid-19 and most other viruses, actually can insert its genetic material into the human genome, thus making its DNA indistinguishable for host human DNA. Thus, to date, only two people under very unusual circumstances have been declared fully clear of the virus (44). Most patients use a cocktail of drugs that interfere with the action of the viral enzymes to prevent the virus from being active in its human host (45), but the viral DNA remains in the host DNA and means that the virus remains present. Most viruses, like the ones behind the flu, the common cold, and the coronavirus, rely solely on their RNA in the cell and once destroyed leave no trace of their former presence in their human host's DNA (46).

The subject of cancer deserves a book until itself, but as I mentioned from the get-go, there is no attempt to be complete here. Cancer often involves a single base mutation in DNA leading to uncontrolled reproduction and growth of cells (47). The mutation may be caused by environmental factors such as carcinogenic chemicals; chemicals alone are not the only cause of cancer (heredity, for example). There are 3 billion DNA "base pairs" or pairs of interacting nucleic acid molecules (48) in the nucleus of a cell. The change in DNA that causes a cancerous growth generally occurs in a gene coding for a regulatory protein, namely those that can be turned on or off to control cellular functions (49). When a change in the DNA leads to a structural change in the resultant protein that makes its structure always in an "on" configuration, then the cell and its progeny will reproduce uncontrollably without the "off" feature normally present to allow for control of cell division and growth (49). As long as the defective DNA is present, the defective protein will be made and the cells will continue to grow out of control and form tumors. Thus, traditional treatments for cancer involve attempts to kill cells such as by chemotherapy or radiation treatment. Today, there is no way to go in and fix the changed DNA; there is hope that in the future such treatments could actually be used (50).

"Homework": if you have found the discussions above to be of interest, I would suggest you read up about the coagulation cascade required to stop our bleeding when we form a scab on a cut. The proteins that stop the bleeding are circulating in the blood but do not clot until exposed externally to air via a cut. This reference is a good place to read up about this process (51). Also, one can look at cyanobacteria that grow in hot springs and make for spectacular colors: their proteins do not denature at high temperature, but rather work best at higher temperatures and can be sluggish at normal physiological temperatures (52). The same is true for enzymes from Dead Sea bacteria—they only work properly in a high salt environment and do not do well in normal saline solutions—the opposite of regular proteins which denature at high salt concentrations (53)

Chapter 2: Chemistry

Chemistry is the study of the materials that make up the world and their reactions and transformations (1). How many different chemicals do we interact with on a given day— thousands, tens of thousands? A typical chemical catalogue used in labs throughout the world has tens of thousands of unique chemicals available for routine sale. Look at all of the materials in your home—wood, glass, steel, rubber, plastic, stone, organic sealants, ceramics, aluminum, paint and more. Look at the vast variety of food in your home—each with tens to hundreds of unique chemicals present. A "simple" apple has 300 unique substances present in it (2). What is amazing is that all of the "things" in the world come from about 100 unique *elements*. Below is a standard "periodic table" of elements first described by Dmitri Mendeleev in 1869 (3) which lists all of the elements known. The elements are arranged in columns and rows according to certain properties. While there are about 100 unique entries, most of the chemicals in the world by mass come from combinations of a fairly small set of those elements. An element is something that cannot be broken down into two different materials. An *atom* is a single unit of an element. An atom, when broken down, is no longer representative of a given an element but rather becomes a basket of subatomic particles. A "compound" is a material made of two or more elements. Carbon dioxide which we exhale in every breath, CO_2, is a compound of two elements containing one atom of carbon and two atoms of oxygen. There is one carbon atom for every two oxygen atoms in a single molecule of carbon dioxide. Hydrogen gas, H_2, is a compound made of two identical hydrogen atoms. Hydrogen gas is extremely flammable, while carbon dioxide is used to put out fires.

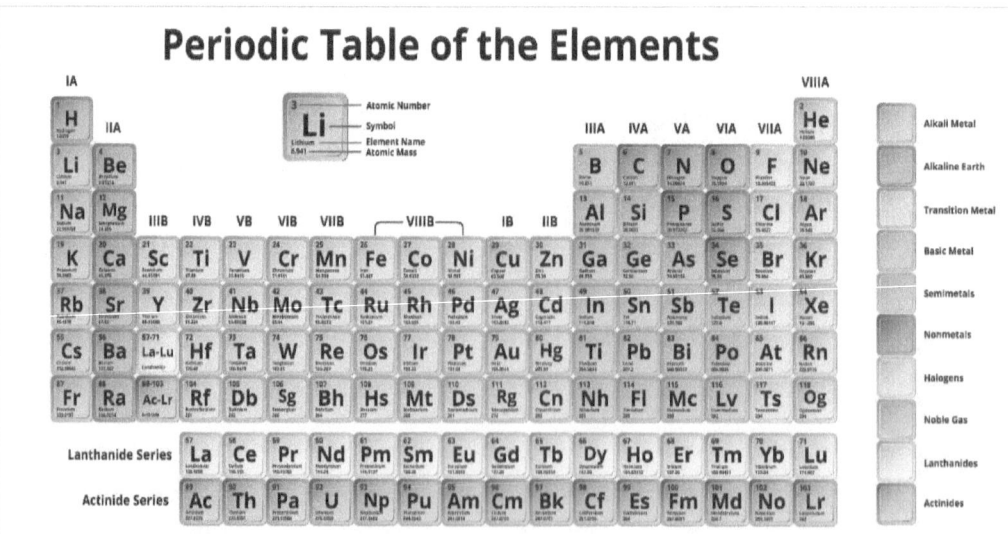

Each square in the periodic table above represents a unique element designated by one or two letters. H in the far left-hand corner represents hydrogen. O is oxygen, Cl is chlorine, etc. Some elements seem easy to figure out such as C for carbon and N for nitrogen. Others may come from French and are thus not obvious such as Au for gold and Pb for lead.

An atom, whatever atom, is composed of two and generally three components for the purposes of this book which does not wish to get into too much technical detail about sub-atomic particles:

Proton, a positively charged species located in at atom's nucleus having an assigned mass unit of 1;

Neutron, an uncharged species and when present, it is present in the atomic nucleus and also has an assigned mass unit of 1.

Electron, a negatively-charged species that is located outside of the nucleus in what is called an orbital and whose mass, relative to the previous particles, is almost nothing.

Below is a schematic view of an atom of carbon. The protons (6) and neutrons (6) are in the nucleus while the electrons (6) are whizzing around on the outside of the nucleus in two distinct orbital types:

CARBON ATOM

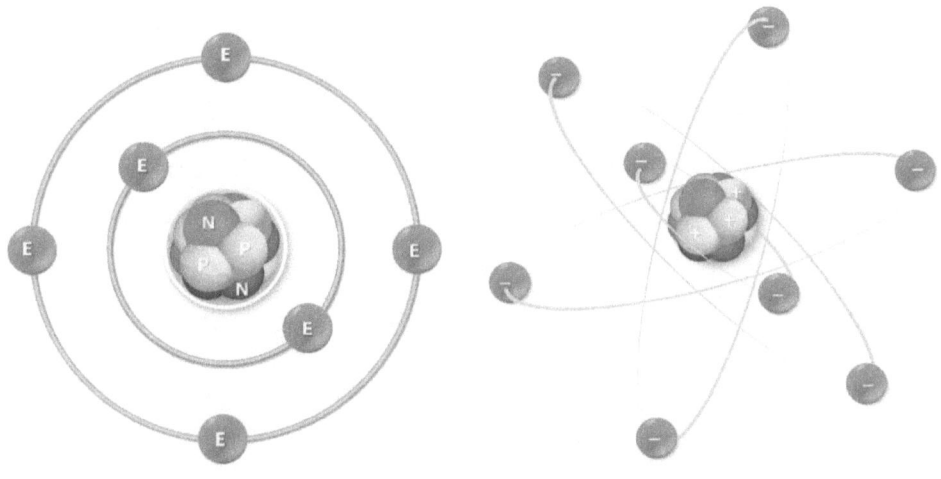

Protons and neutrons have a mass fixed as 1 unit. The number of protons determines *what* element you have. As you can see in the periodic table, H, hydrogen, has the number one. It has one proton. Go further to the right and you can see that carbon has six, meaning six protons (and six electrons to be neutral), nitrogen 7, oxygen eight and so on. Now just think about this for a moment. By adding a single proton (and electron to keep the atom neutral), we go from nitrogen which we cannot breathe and would actually asphyxiate a person to oxygen, which is the gas we need in order to live. Materials with wildly different properties—some are hard, while others are soft, some are solid, while others are gas—differ only in the number of protons/electrons and neutrons present in their atoms. The hard metal titanium and the lighter-than-air gas helium differ only in the number of protons, neutrons, and electrons present in their respective atoms. There are no additional subatomic particles to make them behave differently, only a different number of the same things. There is no special material added to make titanium metal strong, just more of the same of what lighter-than-air helium has.

In the periodic table above, you can see that the atomic numbers progress from left to right and then go down a row and continue. This arrangement allows for elements having similar properties to be lined up in the same column. The similarity of behavior has to do with electron behavior but will not be covered here. For example, the elements in the first column on the left are all reactive metals. The elements in the last column on the right are all inert gases.

The mass of an atom is calculated by the number of protons and neutrons in the nucleus. Electrons are so light that are often ignored when generally calculating the mass of an atom. Atoms of the same number of protons but having different numbers of neutrons are called *isotopes*. You may have heard this word with respect to uranium .([4]) The major isotope of uranium, with mass of 238, does not undergo nuclear fission very easily. Uranium 238 has 92 protons and 238-92 = 146 neutrons in order reach the mass of 238. The isotope 235 (which has three fewer neutrons but obviously the same number of protons in order to still be uranium) is less stable and is used for nuclear fuel and weapons. Another example of isotopes can be seen with carbon:

Carbon 12 has six protons and 6 neutrons and is 98.9% of the carbon in the world

Carbon 13 has six protons (like any carbon atom) and 7 neutrons, is heavier than Carbon 12 and makes up 1.1% of the carbon in the world.

Carbon 14 has six protons and 8 neutrons which leads to instability and radioactive decay. It is found in trace amounts, well less than 0.1% of carbon present. It can be measured and quantified based on its known radioactive decay properties.

All carbon samples should have a mix of all three isotopes unless an effort was made to enrich or remove one of the isotopes. "Enrichment" of uranium 235 is a complex process but can increase the amount of "U235" from its natural 0.72% level to over 97% for bomb-grade material. The remaining "depleted" (of U235) U238 is used for military applications where a very dense metal is desired. The attack plane A-10 uses "depleted" uranium for its armor-piercing shells, for example due to its high density (5).

When one talks of uranium, it is not uncommon to mention plutonium, another nuclear fuel. Plutonium has 94 protons and is the highest atomic number element ever found in nature (6). Anything higher was produced in a lab and may have a half-life of less than a second. Plutonium has been produced synthetically from uranium by the tons for its use in nuclear weapons or as a by-product of nuclear reactors. It was even once used in pace makers but has been replaced by lithium-based batteries. The plutonium produced by the Hanford nuclear site and used in the first atomic explosion was valued at $350 million in 1945 dollars (total expenses until production of first batch of bomb-grade plutonium) (7). Below is a photo of highly enriched uranium.

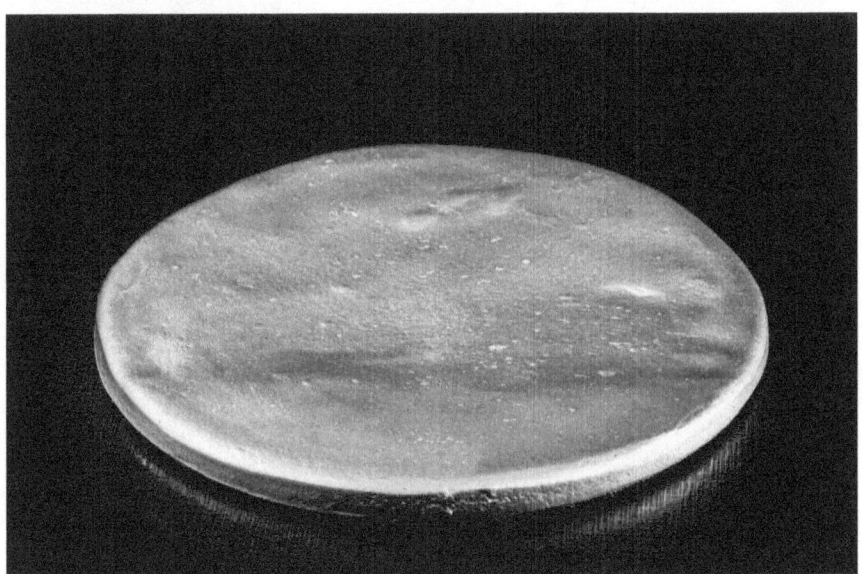

To better understand the masses of atoms and compounds, let's use for an example something we run into every day: table salt, NaCl (sodium chloride). The mass of sodium is 23, based on 11 protons (see periodic table above) and 12 neutrons, while the mass of chlorine is 35, 17 protons and 18 neutrons. Based on these numbers the predicted mass (of one mole, a standard measurement of amount in chemistry) of sodium chloride should be 58. In practice, its measured mass is 58.44 grams. The difference between the calculated mass based solely on protons and neutrons and the actual mass measured will not be discussed here but is explained elsewhere (8).

An atom has equal numbers of protons and electrons to make for charge neutrality. When you hold a gold bar, you are ostensibly holding an uncharged material in which the total number of protons and electrons is equal. Electrons, which circle around the nucleus can at times be added or removed from atoms, sometimes quite easily. As atoms get bigger, the distance between the positive nucleus and the outer orbital negative electrons grows. These electrons are often free to move out of their respective atoms. When we turn on electricity, we are moving electrons that get free of their metallic atoms and respond to applied voltage to make an electrical current. Metals are good at this, so electrical wires are generally made of metals like copper or silver. Wood or glass wires would not allow for electrons to move under voltage and would not work for electrical transfer. The electrical system includes a voltage (110V or 220V generally) that drives electrons through a circuit at a certain current. The moving electrons have energy and allow for work to be done such as lighting a lightbulb or heating an electric burner to cook. Electrons can also be transferred through contact or rubbing between dissimilar materials. Walk along a plush carpet in the winter and you receive electrons from the carpet. Grab a metal door handle and you transfer those excess electrons to the metal and you feel the resulting spark. Triboelectric charge transfer occurs when nonidentical materials transfer electrons between each other via contact, like the carpet to you and you to the door handle (9). This leads to the phenomenon of static electricity which can be cute, annoying or potentially dangerous in the presence of flammable materials (10). Lightning is a transfer of electrons from storm clouds to the Earth, and the thunder is the sound generated by the heating of air around the lightning and its resulting movement at speeds greater than the speed of sound (11). As is well-known, light moves much faster than sound, so we see lightning before we hear the thunder; the closer the storm is to the observer, the less time between the two phenomena.

Atom: neutral, equal numbers of protons and electrons.

Ion: an atom which has had electrons removed (cation) to become positive or added (anion) to become negative. Protons and neutrons in the atomic nucleus remain unchanged.

Where do we meet ions every day? We again return to salt. Most know that table salt is sodium chloride. But it is not sodium and chlorine atoms. The actual salt structure is positive sodium (Na+) and negatively-charged chlorine (chloride, Cl-). Sodium metal is highly reactive and if thrown into water can cause an explosion (12). I had a high school chemistry teacher who had to get rid of the contents of an old lab. There were big chunks of sodium metal under oil. He took them in a rowboat to the middle of a local lake and threw the metal into the lake, with enormous release of energy. Chlorine gas is equally reactive and is used to kill bacteria in our drinking water and swimming pools (13). It was used in World War I for gas warfare. The salt of sodium chloride though is safe and in crystal form pretty inert. The grains of salt that we can see and hold actually contain huge numbers of sodium and chloride *ions* in a crystal lattice. Below is a schematic view of the crystal lattice of table salt. Note the relative size of the ions.

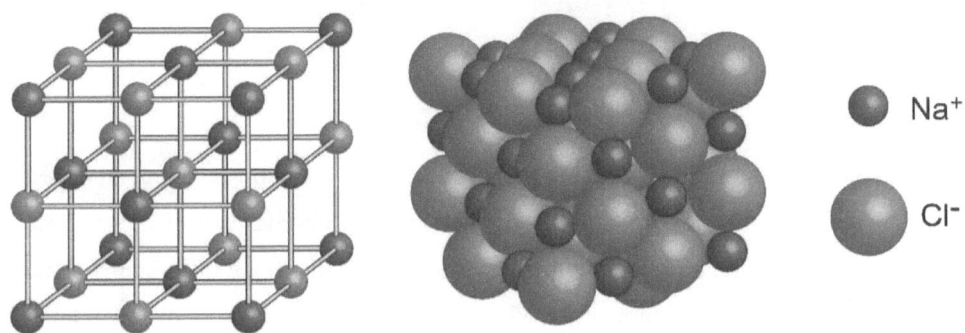

Na+

Cl-

When salt is added to water, the crystals fall apart and free sodium and chloride *ions* are dissolved in the water that stabilizes the individual ions. One can add so much salt to water that it can no longer dissolve any more, but to do so, one has to add a lot of salt. And if one heats the water, more salt dissolves in water than at room temperature. If you put lots of salt into water and then touch the side of the container, it may feel cold. Dissolving salt is called an endothermic process (14), where energy is taken from the surroundings to help dissolve the salt. Other materials when dissolved give off energy; these reactions are called exothermic (14) because they give off energy. Ions also play an important role in electricity. A battery can use ions as "electrolyte" to complement the motion of metallic electrons to allow for flow of electricity in an electric circuit (15). In modern batteries, the electrolyte is often in the form of a gel. Below is a cutaway of a typical 3-volt battery used for electronic devices.

Dry cell battery

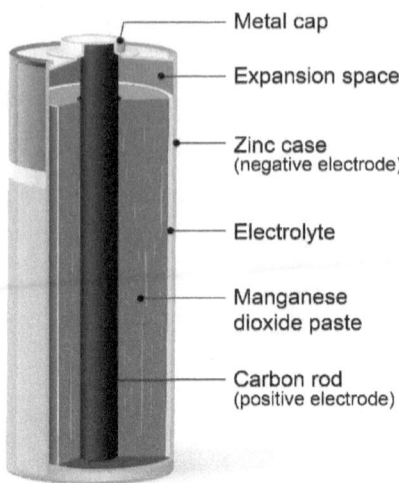

Metal cap

Expansion space

Zinc case
(negative electrode)

Electrolyte

Manganese
dioxide paste

Carbon rod
(positive electrode)

While heating water allows for dissolving more solids, it has the opposite effect on gases: the colder it is, the more one can dissolve (16). Thus, Coca Cola makes its famous drink and bottles it at a cold temperature, to maximize carbon dioxide dissolved in its soda. When we open a bottle of Coke at room temperature, what happens? Gas comes blowing out of the bottle and we can see gas bubbles leaving solution and bubbling upwards. At room temperature, CO_2 is less soluble than it was at the temperature at which Coke was produced and bottled and thus is "super-saturated" and wants out (17). Let the bottle sit open long enough, and the drink goes flat, as much of the gas escapes to the environment and little is left dissolved in the drink.

Whereas salt breaks down into ions when dissolved in water, regular sugar (glucose, $C_6H_{12}O_6$) does not break down in water. The sugar molecules remain intact, but the crystal holding the molecules together is "hydrated" by surrounding water molecules and falls apart (18). Essentially, water interacts with the sugar molecules, and they give up their interactions between themselves that held them in a crystal form that was granulated sugar from the store. Sugar as a solid is the same as sugar in solution. Sugar is a compound made from 6 carbon atoms, 12 hydrogen atoms and six oxygen atoms. Below is a model of sugar:

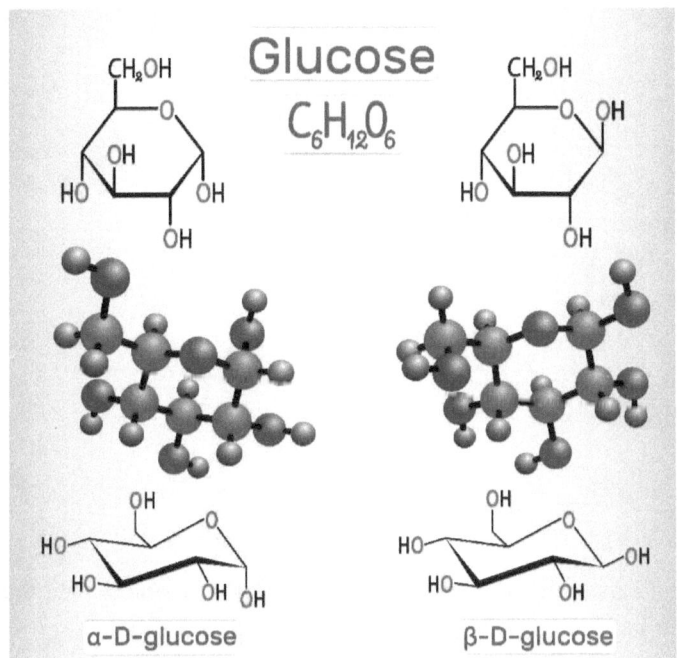

Glucose

$C_6H_{12}O_6$

α-D-glucose

β-D-glucose

Notice in the sugar molecule shown above that the atoms are shown bonded to one another. Whereas water can break the charge interactions between sodium and chloride ions, water does not break the bonds formed of shared electrons connecting carbon to carbon, carbon to oxygen, and oxygen to hydrogen, as these chemical bonds are stable and require a great deal of energy to break. If you heated a solution of either salt or sugar and removed the water, you would get your starting material back again, namely crystals of either salt or sugar.

From the relatively small number of elements in the periodic table, you can build all of the chemicals and molecules in the universe. When you look out the window or take a break in your garden, what is the one natural color that stands out more than any other? Green! The green color is from the chemical chlorophyll, one of the most abundant organic (a compound based on carbon) materials in the world (19). Below is a chemical model of chlorophyll. As you can see, chlorophyll is a very complex molecule, with lots of atoms (every point where lines meet is a carbon atom—this is how chemists write things in a form of shorthand). The Mg^{2+} at the left is a positive magnesium ion (magnesium atom missing two electrons) coordinated to four nitrogen atoms as shown. Chlorophyll has a very important function, namely converting energy from sunlight into chemical energy via photosynthesis (20). The reason it is green in color is that it does not absorb green light. Any color we see is the color not absorbed by the material but rather reflected away towards our eyes (21).

Chlorophyll a

White materials reflects all light, while black color absorbs all colors (and thus black colored surfaces heat up in the summer, whereas white ones generally do not, 22). Have you noticed that in the Fall many tree leaves go from green to red, gold and yellow? For us, it's a symphony of color but for the trees, the pigments (color molecules) that replace or supplement chlorophyll are better suited to get more of the Fall sunlight energy with its changed wavelengths of light (23). This change in color allows trees to wring out more energy before the winter. For us, it's a good reason for a Fall walk in the forest:

When plants were grown in blue or red light only, they grew well, as they could absorb the energy of the light waves. When the same plants were grown in green light, the light was reflected from the leaves and the plants did not grow as well (24).

Let's take a look at some specific elements in the Periodic Table that are found around the house to better understand what roles they play in our daily lives.

Aluminum Al

Aluminum is all around us: aluminum foil, the airplanes in which we fly are made primarily of aluminum alloys, metal window frames in our homes, soft drink cans and so much more. Aluminum is a metal. Metals like aluminum, copper, magnesium and others have certain properties, such as the ability to transfer electricity (think of those copper wires in your house) and heat—like a frying pan (25). Aluminum is found in nature as bauxite which is a salt of aluminum (26). Until efficient methods were developed to prepare aluminum metal from rock via electricity, it was a very rare metal and thus very expensive. If you look at the very top of the Washington Monument, you will see a reflective triangle—made of aluminum metal, which at that time of the Monument's construction was very expensive (27). Now we wrap our sandwiches and leftovers in it and buy 200 square feet of its foil for a few dollars.

Aluminum as bauxite as it occurs in nature and after processing into aluminum metal ingots.

Aluminum's use in transportation and architecture is generally based on two key properties: strength and light weight. The amount of fuel a plane uses is directly related to the total weight of the plane and its contents—the lighter the plane, the less fuel required (28). For that reason, Boeing (and later Airbus) developed a plane primarily made of carbon fiber which has the strength of steel but weighs much less than metals. The Boeing 787 is about 20% more fuel efficient than similar planes made from aluminum alloys (29). On the other end of flight, the SR-71 "Blackbird" spy plane would fly so fast that aluminum would have melted from the high temperatures generated from friction with the air at very high speeds (30). So, the Blackbird was made primarily of titanium, which also is a very strong, light metal but having much better strength at the higher temperatures reached during Mach 3 flight (three times the speed of sound, around 2,000 miles per hour). Since the Blackbird got so hot, no material existed that could completely seal the fuel tanks, so fuel just dripped from the plane onto the runway prior to flight (31). When the plane took off with half a tank of gas, the hot temperatures of the metal allowed the plane to close its fuel tanks by the expanding metal. The plane would then undergo aerial refueling. This is one of the properties of metals and other materials: they expand when heated (32). The Concorde also experienced expansion (of one meter!) at high speeds/temperatures, and on its last flights, the cabin crews inserted their hats into the openings created between equipment panels during supersonic travel. As the plane cooled down, the metal contracted, and those hats are stuck there forever (33). There are some versions of the plane in various museums in which the hats are there to this day.

Carbon, C

Carbon is found in at least ten million unique compounds known to chemists (34). When carbon is present in a molecule, the molecule is referred to as "organic". Carbon is present in all of the critical foods and metabolites used from bacteria to humans. Carbon has unique binding properties that afford it great versatility in its interaction with other carbon atoms as well as with other elements. Carbon alone has many forms, some of the most famous being coal, graphite, and diamond; these are known as allotropes of carbon (35). Synthetic diamonds are made in one method by applying high temperature and pressure to graphite to cause a change in crystal structure to allow the transition from one arrangement of carbon atoms to another (36). It's hard to believe that a diamond ring is from the same family as a lump of coal, but chemically it's true. Carbon is an incredibly important industrial material. Coal and hydrocarbons (like natural gas, propane, gasoline, kerosene, etc.) are the major sources of energy used by humankind (37). Carbon added to iron affords us steel with its important applications in all facets of our lives. As mentioned above, carbon fiber is used in airplanes as well as in race cars to give strength at lower weight (38). Human energy sources such as sugars, fats, and proteins are all based on carbon chemistry. If one wants to learn more about the reactions of carbon, he/she might want to look at an organic chemistry textbook.

From the same material: Diamond & Coal

Helium, He

Anyone who has seen the Macy's Thanksgiving Day Parade knows that helium is lighter than air (39). As air is composed primarily of nitrogen and oxygen gas molecules and helium has a lower density (mass in a given volume) than these two other gases, helium rises (like olive oil over water). Hydrogen gas has the same property, but as hydrogen is highly flammable while helium is inert, zeppelins and floats moved to helium after the Hindenburg disaster in New Jersey in 1937. There is a finite amount of helium in the world and its synthesis from hydrogen is in the realm of nuclear fusion and thus not commercially relevant (40). Thus, helium is often recovered from the large floats for re-use. And as most children know, if you swallow some helium and then speak, you have a funny, high-pitched voice. This outcome results from the changed atmosphere in which the sound waves move (they move more quickly) after their generation in the vocal cords (41).

All of us have either purchased or received a nice, shiny helium balloon for a birthday, anniversary or the like. What happens? Initially, the balloon wants to escape skyward but after a while it loses some of its gas and slowly comes down until it finally sits on the floor. What happened? While the foil balloons look solid, they have a certain permeability to gases, meaning that gases can diffuse out through holes (which we cannot see with our eyes) in the material (42). The release is slow but all helium that leaves goes out into the atmosphere, never to return. Gas that enters? Regular air. So, the balloon slowly surrenders its helium and in return gets some air to come inside, and thus the balloon loses its ability to float.

Silicon, Si

Silicon (43) can have unique electrical properties somewhere between electrical conductors like copper and electrically-insulating materials such as ceramics and glass. Semiconductors (44) play critical roles in all of our electronic devices such as smartphones, computers, modern cars, kitchen appliances and much more. Silicon-based chips are used in virtually every modern device and appliance, while other semiconducting materials such as those based on gallium (see atomic number 31 in periodic table above) are used in LED's used in lighting (45). Below is a picture of a crystalline silicon wafer being constructed into a plurality of chips for use in numerous electronic devices. While other materials have semiconducting properties, silicon is the workhorse for semiconducting electronic chips and the like. The Nobel Prize in Chemistry in 2002 was awarded to researchers who invented semiconductors from organic materials. These materials are used in organic LED's (OLED's) which you can find in bendable screens on cellphones and other devices (46).

Other Elements

Not discussing the tens of other unique elements and their roles in the world around us is not from a lack of respect! For each element, one could write a book or multiple books. Magnesium for example, is used in ladders (47) due to its light weight; additionally, it is worn by workers at Fort Knox over their shoes (48), so that should a bar of gold fall, it won't break their toes. Inert argon is used in fire suppression systems where rare books or important documents are present and water damage would be unacceptable (49). This book is meant to be an introduction and not the final word. We live in an age of unparalleled access to information, so if you want to understand why there would be lead in your water pipes or why gold never tarnishes or why sulfur is so smelly, I encourage you to look it up on the internet.

Each material has its unique properties and with them, its place in the world around us. Many of us benefit from electricity made from nuclear power, driven by the fission properties of uranium atoms. Very few other elements could provide the same amount of energy in a relatively small volume. When I hear of new nuclear subs that don't need to refuel for 25-30 years (50), I can only wish the same was true for my car. When I did my doctorate in Madison, it involved a lot of x-ray crystallography. I remember the x-ray generator having a "beryllium window" which allowed for greater transmission of x-rays (51). Beryllium is #4 in the Periodic Table above. Many "rare earth metals" play critical roles in modern electronics (52). When the Nazis could no longer get certain metals like chromium they needed for their revolutionary jet engines, they had to make do with what they had; the result was that their engines burned out after a very short period of use (53). The US stealth fighter, F-22, benefits from unique ceramic materials that allow the hot gas flow from the jet engines to be directed onto this material when the airplane engages its square nozzles in "thrust vectoring", giving it unparalleled maneuverability (54). Development of such unique materials took years, but they came from the same toolbox—the Periodic Table.

While elements have unique properties with respect to heat/cold, electrical transfer, weight and strength, most things we come into contact in life are *compounds* made from multiple elements. As mentioned previously, sugar is made from carbon, oxygen and hydrogen. All of our foods include compounds composed of multiple different elements bonded together. Amino acids, the building blocks of proteins such as those found in meat, fish, and foul, are composed minimally of carbon, oxygen, hydrogen and nitrogen. Fats are primary carbon and hydrogen, though they have some oxygen as well. Metallic ions such as zinc and copper are important at proper concentrations in our diet, as certain enzymes require them for proper function (55). And of course, vitamins are needed for health, as they are precursors generally for compounds that allow enzymes in our cells to perform properly (56). Below is a schematic of the heme molecule found in four copies per molecule of the blood protein hemoglobin, the red molecules in our blood:

heme B

At the heart of the hemoglobin protein molecule is this heme group that features an iron ion (cation) in the middle. That is the reason why blood has a metallic taste to it, when one cuts his/her lip. The iron plays a very important role that is modulated and controlled by the very large hemoglobin molecule around it. When blood is sent from the heart to the lungs, the high concentration of oxygen there causes oxygen gas (O_2) to bind to the iron atom (57). Simultaneously, a carbon dioxide molecule (CO_2) previously bound to the iron falls off due to the higher pH of the lungs. The carbon dioxide is released as we exhale, while the oxygen-rich blood is returned to the heart to be sent throughout the body to provide critical oxygen to active cells. Hemoglobin with oxygen is red, and yes, hemoglobin with carbon dioxide is blue, as one can see from the veins seen through our skin. When hemoglobin in the blood gets to the cells of the body, the process is reversed: the low pH of the metabolizing cell causes oxygen to jump off of the iron and into the cell where it is needed for proper function in routine energy metabolism. CO_2 produced by the same energy metabolism jumps on the iron for the ride back up to the lungs for release. And all of this occurs without any intervention or thought about it. The binding of one molecule of oxygen to hemoglobin aids in the binding of the next and so on, as each hemoglobin has four heme groups and can carry 4 oxygen molecules at a time. So on a chemical level, hemoglobin is a large protein made primarily of carbon, hydrogen, nitrogen, and oxygen that binds a "cofactor"—heme—made up primarily of carbon and hydrogen and critically an iron ion, and the major function of this ball of chemicals is to successfully move oxygen to respiring cells and remove waste carbon dioxide from those cells and get them out of the body. Pretty impressive what these chemicals can do, no?

HUMAN HEMOGLOBIN

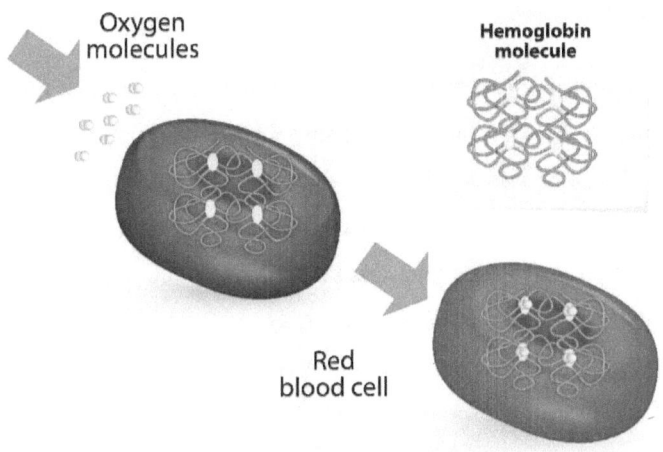

When I did my doctorate in Madison, I used to look out of the lab window on a frozen Lake Mendota, where little huts and tents housed ice fishermen. Almost every winter, there was a report of people dying in one of those structures due to a heater giving off CO, carbon monoxide in a tightly-sealed cabin. Carbon monoxide has one less oxygen atom than carbon dioxide and binds strongly to hemoglobin and does not come off. People are asphyxiated due to a lack of oxygen being transported by hemoglobin throughout the body (58). What's the difference of one oxygen or two? In the case of carbon monoxide and carbon dioxide, the difference is reflected in life or death.

So, as you move through your busy day, take a moment to think about all of the materials that populate our world—marble for kitchen counters, glass for windows, rubber for tires, asphalt for roads, and on and on. Each material is made of molecules or elements, which in turn are made solely of protons, neutrons, and electrons. Now think how absolutely useless marble would be for tires or rubber for stove tops, etc. and then you will appreciate how amazing the materials in our lives truly are and how their properties allow us to live our lives as we do.

There are many experiments that one can do to better understand the properties of material around us. One should look for simple experiments that can be performed safely with things at home or that are easily purchasable. I have found nothing sparks the enthusiasm and curiosity of children as much as doing with them simple scientific experiments. One has to be careful that all safety considerations be taken into account. If one does an experiment that causes the generation of gas—say putting hydrogen peroxide into a contact lens case--he/she must be certain that there is enough ventilation to allow for gas release in order to prevent an explosion.

Chemistry in the kitchen. I once covered a wet salad in a stainless-steel bowl with aluminum foil. In the morning, I took the salad out and saw that the aluminum had changed color and had lots of small holes. I had unintentionally set up a galvanic corrosion cell: two dissimilar metals (to create a voltage) and an electrolyte, the liquid in the salad. Aluminum metal released electrons to the "solution" and aluminum ions fell into the salad, with resulting pinholes in the foil. This is a form of corrosion which we generally associate more with rust on iron-based metals like those used on cars than with aluminum surfaces. Needless to say, that I threw out the salad and started using plastic wrap for metal containers afterwards.

Pitted Aluminum Foil After Voltaic Corrosion

I would be remiss if I finished this section without mentioning water. When I was a graduate student in Madison and acted as a teaching assistant for Biochemistry 201, I gave my first ever lecture to a full class on the properties of water. Most—but not all—materials can experience three phases (60): solid, liquid, and gas. The transition from solid to liquid or from liquid to gas generally occurs at very specific temperatures. For water, at zero degrees Celsius, ice melts into liquid water and at 100°C, water boils and becomes gaseous steam (I am assuming normal atmospheric pressure; at higher locations or under vacuum, these temperatures can be much different). Below is a graph for water as a function of temperature. Note that at 4°C, there is a "triple point" (60), namely the temperature where all three forms of water co-exist.

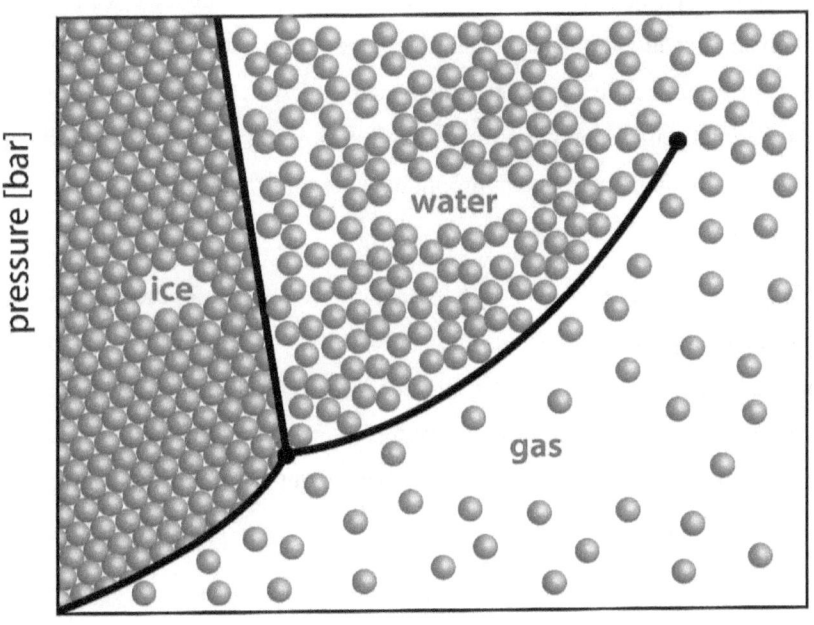

The figure above explains why materials make transitions. Solid ice is very ordered in a crystal lattice—the molecules are limited in how much they can move due to their interactions with other molecules in the crystal. As one adds energy (moving to the right on the X-axis, temperature going up), the molecules, having more energy, can move more freely and eventually they break their interactions and move to a more energetic and disordered state, namely water. If one continues to add energy, molecules can escape from the water into the atmosphere and you have even greater disorder—steam, which is the gaseous phase of water. As I mentioned, not all materials have 3 phases. Carbon dioxide generally goes straight from

37

dry ice (solid) to carbon dioxide gas. Under pressures well over normal atmospheric pressure, a liquid phase can be isolated but not under normal conditions (61).

As is well-known, water is H_2O. There is a phenomenon of "hydrogen bonding" wherein a hydrogen atom of one molecule of water can form a weak interaction with the unpaired oxygen electrons of another water molecule (62). The result is a huge web of these weak interactions between enormous numbers of water molecules. What are the implications? Have you ever filled a cup over its top and the water stood there and did not spill? This behavior is due to the capillary action of water, which is an outcome of hydrogen bonding. And while most materials have their solid version heavier than their liquid counterpart, solid water—ice—floats and because of this property, fish are not killed off in water with sinking ice crushing them. As is well-known, our bodies are about 60% water, so water is truly the liquid of life (63). Many organic liquids are immiscible with water, and many of them will float on water—like oil. Some of the oil actually is dissolved in the water, but most sits above the water due to the oil having a lower density than water and hydrophobic (literally, "water hating") materials do not mix.

As we mentioned temperatures above, I'll note that on the upper end, there is no limit to high temperatures—thermonuclear explosions can generate temperatures in the tens of millions of degrees Centigrade (Celsius) (64). On the other end, there is a limit: -273.15 degrees Celsius is known as absolute zero (65). At this temperature, all atomic motion comes to an end. There is no lower temperature than this value. In practical terms, the lowest temperature ever measured was within 0.1 degree of this value (65).

"Homework": Soaps and detergents are from a class of chemicals known as amphipathic, having both hydrophobic and "hydrophilic" (literally, water loving) parts (66). They are as such so that the hydrophobic part can get into the dirt while the hydrophilic portion removes the detergents/soap and the dirt into the water and out of the clothes, dishes, or hands. For a simple experiment, one can take a plastic disposable cup and fill halfway with water. Swirl it. If the water and cup are clean, nothing or very little should be stuck to the cup above the waterline. Now take a few sips and swirl again. You now see lots of little drops of water on the cup. Why? The chemicals from your mouth include things that have hydrophobic portions that rapidly bind to the cup, but since they are also water-loving, water binds to them in place. One can detect contaminants in water by this behavior (67). Additionally, one can take a similar cup and put in water and oil. The two will remain separate. Now add a highly-colored detergent. Swirl gently with a spoon. You will see that the detergent enters both the water and oil phases as it has parts that feel at home in either. The soap may cause the phases to combine int a single phase and thus become an emulsion (68).

Chapter 3: Physics

Physics may generally be described as the study of the laws of matter and energy (1). Think of gravity for example—gravity relates to the forces between a large object, like the Earth, and a small object like us. The study of physics can apply to the forces between subatomic particles to the motional behavior of distant galaxies.

But before we really get into the subject matter of physics, let's just look at one particular issue, namely how science works. Science, like sports, has rules. Data are generated and a theory is proposed to explain the data—why we got the results we got. If a theory explains the data well, it can remain in place for years, decades or centuries. But once a theory fails to explain phenomena, it must be altered or replaced. Sometimes, it only needs a slight tweak; other times, it requires the wrecking ball. Nothing better illustrates such an idea than the accepted idea based on human observation of the sun going around the Earth then being replaced by a Copernican view of the Earth going around the sun and as of today the Einsteinian concept of relativity that different observers might come to different conclusions as to who goes around whom. But one need not go back so many years. For many years, some experts believed that one could not safely fly faster than the speed of sound (2). When Chuck Yeager succeeded in doing exactly that (2) obviously the theories of transonic flight had to be changed because Yeager's flight—the data—showed that those theories were wrong. If a better theory cannot be offered—better meaning that it more accurately and reliably explains data—then we hold onto the theory we have. But under no circumstances in science should a theory be considered final. If an unknown researcher from a small institution brings the data that support a theory completely at odds with the prevailing thinking, then we go with his/her idea. There is no such thing as "settled science": we are always looking for truth, and if a theory cannot explain ALL of the relevant data, then we will look for a modification; and if a tweak does not help us explain all of the data, then we'll go for a completely new theory. As they taught me in college and graduate school: theories can only be disproven, never proven (3). Science has always been and will always remain a snapshot of how we understand the world around us at this moment. And tomorrow's snapshot will necessarily not be the same as today's.

Physics, like chemistry, is an enormous subject, and as such there is no intention here to go into great detail but rather to highlight some daily or well-known phenomena. Typical introductory physics courses divide the subject matter into mechanics and electricity/magnetism. Mechanics deals with energy as well as the many forces in the physical world. Let's think about friction. If you play air hockey, you know that the plastic puck can travel a great distance with a very gentle push. There is frictional force slowing down the puck from the air itself and from the sides, but if one has had the air turned off at an arcade, he/she knows that the puck does not move much at all when there is no air present to significantly reduce the drag on the puck. Thus, the force you apply to an object can put an object in motion, but forces such as friction remove energy from the object and eventually may bring it to zero motion. The harder you push the puck, the more times it will go back and forth.

Think about holding a ball well over your head. You let go and it bounces back to the level of your head, the next bounce to the height of your chest, then belt, etc. until it barely goes up again. Why? Why shouldn't we have perfect conservation of energy, with the ball jumping back up to its original height—forever? Energy comes in many forms and when the ball hits the ground, some of kinetic energy is converted to sound (also a form of energy), while some of the energy is transferred to the ground in the form of heat or deformation of say the dirt on which it fell while air also takes from the ball's energy. The potential energy of the ball that it had as you held it over your head was the maximum energy in this system. All energy is conserved, but much of it is wasted in sound and heat and friction, with much less available for the main activity, namely the ball bouncing. Many systems, from solar panels to human cells to car engines use only a small percentage of the total energy available to them. Reducing energy inefficiency is a very important part of reducing energy usage and costs.

While biology is what best describes what goes on within us and chemistry describes the material world around us, physics represents, in some ways, the laws governing the activities in the physical world. A study guide for a driver's license is in many ways a physics book: how long it takes the car to stop, what happens when the ground is wet, how much space do you need in front of you when you double your speed, don't use the bright lights in fog, and on and on. Flight is based on air passing more quickly over a wing than under and a resulting lift applied to the wings that takes the entire plane up into the air. The plane's engines work by compressing inlet air, followed by mixing with gas and ignition, with high temperature gas expansion providing thrust to move the plane forward against the drag on the plane. Bridges are built with spans and foundations to allow for maximal physical strength in the face of high winds or monsoons or the like. Watch the original footage of the first moon shot, Apollo 11, and see physics in action—every single activity and the timing of thrusts, docking, landing and the like were all calculated based on known physical phenomena and their associated equations.

The mass of an object in grams or kilograms does not change based on its environment. Its weight though does change as a function of gravity. The gravity on the moon is about $1/6^{th}$ of the gravity on Earth. In space, the gravity is not zero but it is so small that it is generally referred to as microgravity. Everyone has seen pictures of astronauts putting their food in the space in front of them and the food does not fall! Due to the distance between the space craft and Earth, the pull of Earth's gravity is overcome; it took huge amounts of energy to push the rocket or the like out of Earth's gravitational pull and into space. In microgravity, there is a general feeling of weightlessness which allows one to pick up heavy loads that would be impossible for a single person to lift on Earth (4).

The Manhattan Project, in its inception, was definitely a chemistry and physics project (as well as an Army project). There were issues of materials and their properties (chemistry) but also energy and forces (physics). With the beginning of the atomic age, atomic energy and weapons moved more in the direction of physics. Even the part of a bomb or missile that includes the nuclear fuel and detonation mechanism is routinely called the "physics package"(5). Einstein's famous equation, $E = mc^2$, namely the energy released in the conversion of mass to energy is equal to the mass converted times the square of the speed of light (300,000 kilometers per second). Based on this equation, one realizes that the 15 kiloton (15,000 ton) equivalent of TNT that exploded over Hiroshima was the result of 700 milligrams of mass turning into energy (6). To put that into perspective, that mass is less than $1/3^{rd}$ the mass of an American dime. In that bomb, only 0.1% of the uranium underwent fission reactions. Today's thermonuclear weapons are so powerful that even the fission-reluctant U238 undergoes fission due to the high energy of the neutrons produced by fusion in the secondary stage of the bomb (7).

My favorite physics experiment was the "monkey drop experiment". One can see it performed at MIT here (8). The basic concept is that one aims a rifle at a toy monkey hung high and away. When one pulls the trigger, the projectile is fired and the monkey starts to fall simultaneously. The "bullet" and the monkey experience the same force of gravity and thus meet up before the monkey hits the floor. I first saw this experiment in high school. Unfortunately, at Harvard one of our first semester Physics 11a final question involved this experiment. The professor had the bullet and the monkey falling at different gravity values: when would they meet? I kept getting -2 seconds. I was pulling my hair out—I still had hair then. In the end, the right answer was "they will never meet". That was a bummer.

I would be remiss if I did not mention electromagnetism, one of the major fields of physics. When an electric current runs through a wire, a magnetic field is generated by the flow of the electrons. The same is true in reverse: the motion of magnets can create an electrical current (9). It is the flow of water that spins the magnets in the turbines of a hydroelectric generation system that generates *de novo* electricity in the wires of metal coils. The whirling magnets generate an alternating current that supplies us with clean electrical power for all of our needs. As is often the case, one form of energy—falling water—is converted into another form of energy—electricity. High temperature steam produced by nuclear, gas, coal or solar power also drives turbines to generate electricity by a similar mechanism. The conversion of one form of energy to another, more useful, is equally true when we burn gas to drive a mechanical engine. Airplanes use aviation fuel that has a very high energy density so that they have a maximal amount of energy available in the smallest volume possible. There are no gas stations in the clouds.

The turbine room at Hoover Dam: water spinning magnets associated with these turbines drives electricity production followed by transmission.

And one last subject: noise-cancelling headphones. I bought a pair last year and whenever someone in the house wears them, people scream to get his/her attention, but often with little success. How do they work? As we learned above, sound travels in waves (see figure below). As one can see from throwing a stone into water, waves can be additive—one wave can be added to another. This can make the result bigger—or smaller. So, sound-cancelling headphones measure the background noise around you and then generate waves to cancel these noise sound waves. In the figure below, you see an amplitude—the height of the wave. If one makes a negative amplitude at the same wavelength, one can make the two waves add up to zero. Thus, you feel a total silence. These systems cannot cancel all noises, especially if the noise is not constant but something that suddenly appears. Still, one who has used such headphones or earbuds knows the incredible feeling of disconnecting from the surroundings and being in a totally quiet, still space. I met with a company that is working on systems to cancel the noise we hear constantly on planes from the engines.

SOUND WAVES

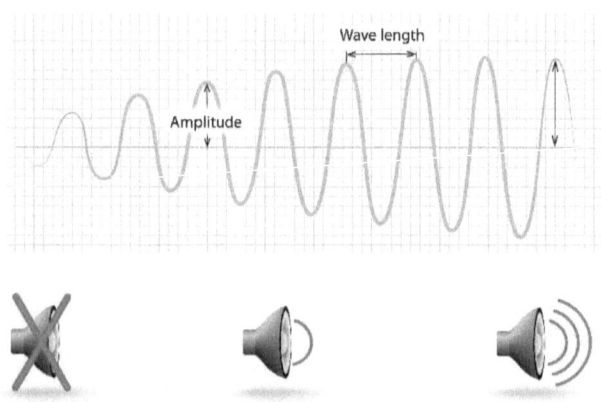

I have been brief in my discussion of physics as there are many who are far more capable of presenting the beauty and intricacies of this field than I can. As I mentioned, I am a biochemist, the son of a chemist. My father taught medicinal chemistry for over forty years. As I pointed out previously, there is an enormous amount of material available on the internet for continued study of general or specific areas in physics or any of the sciences. This pamphlet was never meant to be complete; rather, my goal has been to introduce information and concepts that are well-known but which may not be known to those who do not routinely work or dabble in the sciences. If the subjects of this work have interested you, I would strongly encourage you to keep learning about them. There is no shortage of information available today on any and every subject. When I was a kid, if I wanted to know something, it was either the *Encyclopedia Britannica* we had (from a time when the US only had 48 states…) or a visit to the local library. Today, we have all of that information and much more literally at our fingertips.

Chapter 4: Conclusions

When we make free video calls from one side of the world to the other, we today think nothing of it. But if one were to go back 50, 100 or more years, the idea that you could touch a screen and then speak with a person thousands of miles away in a matter of seconds would have been considered beyond science fiction. The same is true for air travel and space travel and the availability of all human knowledge on a device that fits in a pocket. We live in an extraordinary world, from the amazing properties of materials based on the various combinations of protons, neutrons, and electrons to our own bodies and the billions of miles of DNA that we have within us. One does not have to be a professional scientist to see the amazing beauty of the world in which we live and to pause and let sink in the extraordinary complexity and sublime beauty of the material, living, and physical world of ours.

Endnotes

Biology

1.https://www.scientificamerican.com/article/are-viruses-alive-2004/

2.https://www.ncbi.nlm.nih.gov/pubmed/23829164

3.https://www.sciencefocus.com/the-human-body/how-long-is-your-dna/. Based on 2 meters of DNA per cell and 10 trillion cells with DNA and the average distance between the Earth and the moon to be 384,400 kilometers, all of the DNA in an average person would come to 20 billion kilometers which would allow for 26,014 round-trips to the moon. 10 trillion cells x 2 meters DNA/cell x 1/1000 meters-to-kilometers x 1/384,400 kilometers to moon x 1/2 for round trip = 26,014 round-trips.

4.https://www.everydayhealth.com/news/10-amazing-facts-about-your-blood-vessels/

5https://en.wikipedia.org/wiki/Pulmonary_alveolus

6.https://www.sciencemag.org/news/2018/09/there-are-about-20000-human-genes-so-why-do-scientists-only-study-small-fraction-them

7.https://www.ncbi.nlm.nih.gov/books/NBK22599/

8.A professor in Madison related to me that a student had asked him how many "things" happen in the cell each second. He assumed that she meant enzymatic reactions, so he used the number of cells, an estimated concentration of enzymes and a value for reactions per second to reach this number, which is not based on any experiments. Estimated number of stars in the universe: 10^{24}.

9.Calculation: 80 years x 365 days per year x 24 hours per day x 60 minutes per hour x 72 beats/minute average = 3,027,456,000 beats.

10.https://en.wikipedia.org/wiki/Sperm

11.https://en.wikipedia.org/wiki/Human_fertilization

12.https://en.wikipedia.org/wiki/Cellular_differentiation

13.https://www.ncbi.nlm.nih.gov/pmc/articles/PMC2117903/

14.http://hyperphysics.phy-astr.gsu.edu/hbase/vision/rodcone.html

15. Mr. Philip McCrea, New Trier High School, Biology 104, Summer School 1983.

16. https://www.aao.org/eye-health/tips-prevention/how-humans-see-in-color

17. https://courses.lumenlearning.com/physics/chapter/26-2-vision-correction/

18. http://scienceline.ucsb.edu/getkey.php?key=134

19. http://hyperphysics.phy-astr.gsu.edu/hbase/Sound/earsens.html

20. https://www.science.org.au/curious/people-medicine/how-do-our-tastebuds-work

21. https://www.sciencedaily.com/releases/2006/08/060807121323.htm

22. http://chemistry.elmhurst.edu/vchembook/568denaturation.html

23. https://www.scientificamerican.com/article/why-is-it-that-eating-spi/

24. https://www.ncbi.nlm.nih.gov/pmc/articles/PMC2234081/

25. https://www.livescience.com/66028-why-mint-makes-your-mouth-cool.html

26. https://www.nature.com/news/human-nose-can-detect-1-trillion-odours-1.14904

27. https://pubs.acs.org/doi/10.1021/acs.jafc.8b06183

28. https://www.ncbi.nlm.nih.gov/books/NBK26860/

29. https://en.wikipedia.org/wiki/Antibody

30. https://www.ncbi.nlm.nih.gov/books/NBK27160/

31. https://time.com/5827450/who-coronavirus-antibodies-reinfection/

32. https://www.ncbi.nlm.nih.gov/pmc/articles/PMC6386720/

33. https://en.wikipedia.org/wiki/Transplant_rejection

34. https://www.sciencedirect.com/topics/neuroscience/ciclosporin

35. https://news.harvard.edu/gazette/story/2020/01/jack-strominger-to-retire-after-a-lifetime-of-achievement/

36. https://www.sciencedirect.com/topics/biochemistry-genetics-and-molecular-biology/lac-operon

37. https://www.nature.com/articles/sj.bdj.2017.1048

38.http://homepage.smc.edu/wissmann_paul/physiology/krebscycle.html

39.https://www.health.harvard.edu/staying-healthy/understanding-the-stress-response

40.https://www.britannica.com/science/tropism

41.http://www.plantphysiol.org/content/156/4/1819

42.https://www.researchgate.net/publication/315811492_Tuned_in_plant_roots_use_sound_to_locate_water

43.https://en.wikipedia.org/wiki/HIV

44.https://www.medicalnewstoday.com/articles/2nd-person-cured-of-hiv-thanks-to-stem-cell-transplant

45.https://www.healthline.com/health/hiv-aids/understanding-the-aids-cocktail#drug-classes

46.https://www.immunology.org/public-information/bitesized-immunology/pathogens-and-disease/immune-responses-viruses

47.https://www.cancer.gov/publications/dictionaries/cancer-terms/def/mutation

48.https://www.genome.gov/human-genome-project/Completion-FAQ

49.https://www.ncbi.nlm.nih.gov/books/NBK21513/

50.https://www.sciencemag.org/news/2020/02/cutting-edge-crispr-gene-editing-appears-safe-three-cancer-patients

51.https://www.ncbi.nlm.nih.gov/pmc/articles/PMC4260295/

52.https://en.wikipedia.org/wiki/Thermophile#Thermophile_versus_mesophile

53.https://books.google.co.il/books?id=TlCHxp2MpeoC&pg=PA88&lpg=PA88&dq=dead+sea+enzymes+salt&source=bl&ots=jmwa5pwWGF&sig=ACfU3U0rkiPZ5vWKrNxUEySQZBxiMXNABQ&hl=en&sa=X&ved=2ahUKEwi-t7jsgZHpAhUwzoUKHf7TDc84ChDoATADegQICRAB#v=onepage&q=dead%20sea%20enzymes%20salt&f=false

Chemistry

1.https://www.merriam-webster.com/dictionary/chemistry

2.https://www.basf.com/global/en/media/magazine/archive/issue-6/the-chemistry-of-apples.html

3.https://www.rsc.org/periodic-table/history/about

4.https://www.radioactivity.eu.com/site/pages/Uranium_Isotopes.htm

5.https://www.military.com/dodbuzz/2018/05/14/air-force-debates-replacing-depleted-uranium-rounds-10.html

6.https://en.wikipedia.org/wiki/Plutonium

7.https://www.pbs.org/video/ksps-documentaries-secret-mission-hanford/ see at 47:40 Col. Mathias

8.https://en.wikipedia.org/wiki/Atomic_mass#Relationship_between_atomic_and_molecular_masses

9. https://en.wikipedia.org/wiki/Triboelectric_effect

10.https://www.sumitomo-chem.co.jp/english/rd/report/files/docs/2018E_6.pdf

11.https://www.nationalgeographic.com/environment/natural-disasters/lightning/

12.https://chem.libretexts.org/Bookshelves/Inorganic_Chemistry/Modules_and_Websites_(Inorganic_Chemistry)/Descriptive_Chemistry/Elements_Organized_by_Block/1_s-Block_Elements/Group__1%3A_The_Alkali_Metals/Z%3D011_Chemistry_of_Sodium_(Z%3D11)

13.https://www.bbc.com/news/magazine-27057547

14.https://www.discoveryexpresskids.com/blog/exothermic-vs-endothermic-chemistrys-give-and-take

15.https://batteryuniversity.com/learn/article/bu_307_electrolyte

16.http://scienceline.ucsb.edu/getkey.php?key=2044

17.https://www.scientificamerican.com/article/the-business-of-fizziness-find-your-sodas-fizz/

18.https://courses.lumenlearning.com/cheminter/chapter/dissolving-process/

19.https://books.google.co.il/books?id=XF_kL1hrs3MC&pg=PA161&lpg=PA161&dq=chlorophyll+most+abundant+organic+compound&source=bl&ots=_Z3A6yu6CU&sig=ACfU3U3

6t1H9UmxemITg7fWqDx8TXn1R5g&hl=en&sa=X&ved=2ahUKEwjZgq7M6YXpAhXB16
QKHTlpDEAQ6AEwE3oECAgQAQ#v=onepage&q=chlorophyll%20most%20abundant%2
0organic%20compound&f=false

20.https://www.avasflowers.net/the-function-of-chlorophyll-in-plants

21.https://www.chem.purdue.edu/jmol/cchem/color.html

22.https://scienceline.ucsb.edu/getkey.php?key=3873

23.https://pss.uvm.edu/ppp/articles/fallleaves.html

24. https://www.greenhousetoday.com/does-the-color-of-light-affect-plant-growth/

25. https://www.doitpoms.ac.uk/tlplib/thermal_electrical/printall.php

26.http://www.australianbauxite.com.au/Bauxite-Types-and-Uses

27.http://www.americaslibrary.gov/jb/gilded/jb_gilded_monument_3.html

28.https://en.wikipedia.org/wiki/Fuel_economy_in_aircraft

29.https://www.boeing.com/commercial/787/

30.https://airandspace.si.edu/stories/editorial/10-cool-things-you-may-not-know-about-
museums-lockheed-sr-71-blackbird

31.http://physicsbuzz.physicscentral.com/2011/02/betrayed-by-heat-sr-71-blackbird.html

32. https://www.physlink.com/education/askexperts/ae40.cfm

33.https://www.airliners.net/forum/viewtopic.php?t=224061

34.https://science.jrank.org/pages/1202/Carbon-Why-carbon-special.html

35.https://courses.lumenlearning.com/introchem/chapter/allotropes-of-carbon/

36.https://www.sciencedirect.com/science/article/pii/0022024886902502

37.https://drillingmatters.iadc.org/value-of-hydrocarbons/

38.https://www.tapplastics.com/product_info/why_use_carbon_fiber

39.https://science.howstuffworks.com/helium2.htm

40.https://www.wired.com/2016/06/dire-helium-shortage-vastly-inflated/

41. https://www.mentalfloss.com/article/21590/why-does-inhaling-helium-make-your-voice-sound-funny

42. https://www.thoughtco.com/why-do-helium-balloons-deflate-4101553

43. https://www.electrical4u.com/silicon-semiconductor/

44. https://www.extremetech.com/extreme/208501-what-is-silicon-and-why-are-computer-chips-made-from-it

45. https://www.ledsmagazine.com/manufacturing-services-testing/research development/article/16695262/gan-on-silicon-a-breakthrough-technology-for-led-lighting-magazine

46. https://en.wikipedia.org/wiki/OLED

47. https://www.nytimes.com/1997/03/16/nyregion/safety-first-when-using-a-ladder.html

48. https://books.google.co.il/books?id=ca_MI5tK-YIC&pg=PR14&lpg=PR14&dq=magnesium+fort+knox+shoes&source=bl&ots=j-snC2fn5E&sig=ACfU3U1qQuUl4nTA3BrsHWrBRD1WvbdGjg&hl=en&sa=X&ved=2ahUKEwjp_7u6hobpAhVJ4aQKHQ_KANwQ6AEwEnoECAsQAQ#v=onepage&q=magnesium%20fort%20knox%20shoes&f=false

49. https://www.nist.gov/system/files/documents/el/fire_research/R0000257.pdf

50. https://en.wikipedia.org/wiki/Nuclear_submarine

51. https://chemistry.stackexchange.com/questions/24486/why-is-beryllium-transparent-to-x-rays

52. https://cacm.acm.org/magazines/2019/3/234917-electronics-need-rare-earths/fulltext

53. https://nationalinterest.org/blog/the-buzz/hitlers-jet-fighters-tried-turn-the-tide-world-war-ii-they-25146

54. https://books.google.co.il/books?id=-nrZqzQs3jMC&pg=PA9&lpg=PA9&dq=thrust+vectoring+ceramics+f-22&source=bl&ots=-y1gYNJ82x&sig=ACfU3U0GmAVymd6k_ydUdBm0LdiM3kqDHg&hl=en&sa=X&ved=2ahUKEwjemdbMpobpAhUR3KQKHTgeBPAQ6AEwHXoECAsQAQ#v=onepage&q=thrust%20vectoring%20ceramics%20f-22&f=false

55.https://academic.oup.com/jn/article/130/11/2838/4686136

56.https://chem.libretexts.org/Bookshelves/Introductory_Chemistry/Book%3A_The_Basics_of_GOB_Chemistry_(Ball_et_al.)/18%3A_Amino_Acids%2C_Proteins%2C_and_Enzymes/18.09_Enzyme_Cofactors_and_Vitamins

57.https://themedicalbiochemistrypage.org/hemoglobin-myoglobin.php

58.https://themedicalbiochemistrypage.org/hemoglobin-myoglobin.php

59.https://www.chem.wisc.edu/deptfiles/genchem/netorial/rottosen/tutorial/modules/electrochemistry/03voltaic_cells/18_31.htm

60.https://courses.lumenlearning.com/introchem/chapter/three-states-of-matter/

61.https://en.wikipedia.org/wiki/Liquid_carbon_dioxide

62.https://chemdemos.uoregon.edu/demos/Capillary-Action

63.https://www.usgs.gov/special-topic/water-science-school/science/water-you-water-and-human-body?qt-science_center_objects=0#qt-science_center_objects

64.http://www.atomicarchive.com/Effects/effects7.shtml

65.https://timesofindia.indiatimes.com/home/science/Lowest-temperature-ever-in-universe-recorded-at-Italian-lab/articleshow/44907265.cms

66.https://en.wikipedia.org/wiki/Amphiphile

67. https://patents.google.com/patent/US20170184562?oq=bauer+raisch+water

68. https://en.wikipedia.org/wiki/Emulsion

Physics

1.https://www.merriam-webster.com/dictionary/physics

2.https://www.history.com/this-day-in-history/yeager-breaks-sound-barrier

3.https://www.nsta.org/publications/news/story.aspx?id=52402

4.https://www.nasa.gov/audience/forstudents/5-8/features/nasa-knows/what-is-microgravity-58.html

5.https://en.wikipedia.org/wiki/Nuclear_weapon_design

6.https://www.discovermagazine.com/technology/numbers-nuclear-weapons-from-making-a-bomb-to-making-a-stockpile-to-making

7.https://en.wikipedia.org/wiki/Thermonuclear_weapon

8.https://www.youtube.com/watch?v=cxvsHNRXLjw

9.http://www.mbgnet.net/fresh/rivers/dams.htm